HOARD
L·E·A·D·E·R·S·H·I·P

VOLUME ONE

HOARD'S DAIRYMAN

Published by W.D. Hoard & Sons Company
Publishers since 1870

Copyright © 2015 by *Hoard's Dairyman*
All rights reserved. No part of this book may be reproduced or transmitted in any form or by any means, electronic or mechanical, including photocopying, recording, by any information storage and retrieval system, or otherwise without permission in writing from W.D. Hoard & Sons Company, with the exception of brief excerpts in reviews or as provided by USA copyright law.

Editor — Ali Enerson
Art Director — Ryan Ebert

Printed in the United States of America.

hoards.com

Library of Congress Control Number: 2015902313
Hoard's Dairyman
Hoard Leadership / Hoard's Dairyman
ISBN 978-0-9960753-2-9

19 18 17 16 15   1 2 3 4 5
First Edition

INTRODUCTION

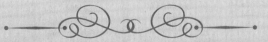

W.D. Hoard's passion for all things dairy took root early in life as he walked past a Munnsville, N.Y., dairy farm each day on his way to and from school. At age 16, after showing great interest in cows, Mr. Waterman Simonds offered Hoard a position caring for his 50-cow herd. As part of the employment agreement, Mr. Simonds insisted that Hoard spend an hour each day reading about improved farming practices in addition to working on the dairy farm. Over the next five years, Hoard emulated the New York dairy farmer whom he called "a great reader and student of the farm." With five years of training under his belt, Hoard turned westward and moved to Wisconsin in 1857.

Entrenched with that knowledge, Hoard later began writing about dairy farming in his fledgling newspaper, the *Jefferson County Union*, following his military service in the Civil War. Hoard advocated that Wisconsin farmers switch from wheat farming to that of dairy, an agricultural endeavor that Hoard believed would better suit the Badger State's fragile, glaciated soils.

From those first columns in 1870, Hoard's dairy crusade rapidly outgrew his weekly country newspaper. In 1885, a national dairy farm magazine was launched, and by 1888, his supporters dubbed him the "Cow Candidate" as he went on to be elected Wisconsin's 16th governor and established the nation's first dairy school at the University of Wisconsin during that term in office.

By 1912, *Hoard's Dairyman* magazine was circulating in every civilized nation on the globe. Copies were sent weekly to China, Japan, South Africa, Russia, India, Australia, New Zealand, Canada and all the countries on the European continent.

Mail bags full of magazines on their way to the post office.

W.D. Hoard & Sons staff in the late 1880's.

On August 25, 1951, *Hoard's Dairyman* editor W.D. Knox began what has now become a 60-plus-year tradition of printing quotes by our founder, W.D. Hoard, on our Editorial Comment page. That page also features an image of his beloved Hoard's Dairyman Farm. During this timespan, some 1,300 unique Hoard quotes have appeared in the magazine alongside the current editor's perspective on our domestic and global dairy industry.

Honored nationally and internationally, and cited as the "father of American dairying" and the "father of alfalfa culture," W.D. Hoard has been quoted by successive generations since his passing in 1918. These unique quotes are as true today as when they were first penned by Hoard nearly a century ago.

In honor of this tradition, the *Hoard's Dairyman* editors carefully selected W.D. Hoard's most famous and pithy observations about dairy farming. As a result, *Hoard's Dairyman* is pleased to present these timeless quotes in a two-part book series titled *Hoard Leadership* and *Hoard Legacy*.

We hope you enjoy reading *Hoard Leadership* and *Hoard Legacy*, which we believe will soon become timeless classics. It embodies W.D. Hoard, who served as Wisconsin's 16th governor, president of the University of Wisconsin Board of Regents, and was named "Wisconsin's Most Distinguished Citizen" at the World's Fair in San Francisco, Calif. Above all, Hoard should be remembered as publisher, farmer and advocate for all things dairy.

Corey A. Geiger
Managing Editor
Hoard's Dairyman

"The American farmer, from the foundation of the government, has constituted the most important class in American society. The agricultural brain has been the parent brain of the nation."

1870 W.D. Hoard begins publishing the *Jefferson County Union* on March 17, 1870, a weekly newspaper for residents of Jefferson County.

1871 — To combat the depletion and erosion of soil resources brought about by a one-crop system of agriculture, W.D. Hoard founded the first dairymen's association in Jefferson County, Wisconsin.

> "There is no vocation on God's green earth that calls for higher elements of character, for deeper research, for grander nobility of nature than that of the farmer."

> "Men forget that a cow needs just as much pure air and clean surroundings for the preservation of her life as men do for theirs."

1872 Hoard led a group known as the "seven wise men" in founding the Wisconsin Dairymen's Association. Through it, the dairy industry became the backbone of a permanent, soil-building and conserving type of agriculture. (Shown right)

1873 W.D. Hoard secured a reduction in freight rates and obtained the first refrigerator car ever to ship Wisconsin cheese to eastern markets.

"Not long since, we heard a farmer remark: 'I wish I had a better education. I would try something else than farming.' The man did not see that what he really needed was education enough to be a good farmer."

1885 On January 23, *Hoard's Dairyman*, "a journal devoted to dairy farming," is founded. (Shown right)

1887 "Original cow census," the very first one carried out in the U.S., was undertaken by Hoard in the town of Ellisburg, N.Y.

"I never saw a poor farmer yet that was a good citizen."

"There is no finer element in American society than the intelligent, well-cultured farmer."

"It cannot be a good thing for a state when the farming population decays."

1888 W.D. Hoard elected 16th governor of the state of Wisconsin.

1890 W.D. Hoard's son, Frank Ward Hoard, was named business manager of the W.D. Hoard and Sons Company.

"There is practically no exhaustion of nature's resources in the product of the cow. It is a regenerative, not a degenerative, wealth that the cow gives us. She is a builder of wealth, of individual and community life, a wonderful food producer without destruction of the original source of supply as in the case of the forest and the mine."

1890 Dairying achieves academic recognition with the establishment of the first dairy school in America at the University of Wisconsin. Hoard is governor.

1890 Steven M. Babcock develops the Babcock test for milkfat at the University of Wisconsin.

"Now, when a farmer can grow a plant like alfalfa, pound for pound, it is worth as much to produce milk as bran, or that 11 pounds of it are the equivalent of 8 pounds of mixed grain; it is a proposition that ought to appeal to the brains of every farmer in the United States."

1890 The tuberculin test is introduced.

1891 — *Hoard's Dairyman* publishes first article on alfalfa. Began intensive promotion. The crop is now the greatest forage-producing plant in the United States. And Hoard is now credited as "the father of alfalfa culture."

"When I came to Wisconsin, all you had to do was to tickle the soil and it would laugh with the harvest. But as we kept on farming, and each man thinking only of what he was going to get out of the soil in the fall of the year, gradually we began to see a depletion of the productive forces. The result has been that the people of today need a great deal more intelligence in regard to the handling of the soil."

1891 L.L. Van Slyke developed formulas for rational payment of milk constituents and cheese yield. Formula still in use today.

1892 Grass or hay silage first recommended as a means of preserving the hay crop from weather damage. Practice now widely followed.

"On the Hoard's Dairyman Farm, we have been growing from 100 to 180 tons of alfalfa a year for 10 years. Yet, we can count scores of farms within sight that have made but very little, if any, attempt at it as yet. Sometimes we think that it requires a ton of proof to put an ounce of conviction into some men's minds."

1895 Began promotion of tuberculosis eradication. A bitter 45-year campaign, with *Hoard's Dairyman* leading the fight to free herds from heavy health losses and to protect the consuming public. Struggle cost thousands of canceled subscriptions.

1899 Hoard purchased a 193-acre farm just north of Fort Atkinson. Provided a place where *Hoard's Dairyman* editors could "keep their feet on the ground."

> "There is no branch of agriculture in which so large a proportion of those engaged are so uniformly prosperous as in dairying. Also there is no system of farming in which the opportunity for greatly increasing the profits is so great as in dairying."

1899 Plant-scale pasteurization proven to kill tuberculosis bacteria. Finding convinces dairy plants to install pasteurization equipment.

1900 Hoard campaigns strongly for the Grout Bill which called for an increase in the tax to 10 cents a pound from 2 cents on oleomargarine colored to resemble butter.

" The prosperity of the city is bound up in the prosperity of the farm."

" How to feed the world is a far more troublesome problem than how we are to consume the fruit of the earth."

" Temperament first decides form, and form governs function."

1905 The first cow testing association organized in Michigan.

1905 Both news and ads appeared on the cover in the early days.

"Every adulteration or counterfeit of a food product is a blow at the farmer himself. Between the farmer and the consumer stands the adulterator, the counterfeiter, the fraudulent mixer of food with substances that, whether harmful or not, are a cheat upon the consumer and an outrage upon the farmer as a producer of food, robbing him of his rightful market."

1906 American Dairy Science Association founded.

1906 Associate Editor A.J. Glover, who joined the Hoard's staff in 1904, serves as general manager of the first National Dairy Show held in Chicago, Ill. Holds post until 1913.

"No man can become successful in any business without constant study of all the points that make up the business. Dairying is not exempt from this rule. The only difference is that a dairyman needs a wider course of study and more sound judgment in the application of the principles, than many professions with a more pretentious name."

1908 After being appointed to the University of Wisconsin Board of Regents, Hoard is selected to serve as Regent president from 1908 to 1911.

1908 The first compulsory pasteurization law (Chicago) for all milk, except that from tuberculin-tested cows, adopted.

"It has been tested and proven that we cannot raise wheat in Wisconsin. But we can raise good butter and good cheese, without killing our land, but rather enriching it."

"The farmer of the future must know more of certain things, chemistry, bacteriology, and mechanics, than did the farmer of the past."

1908 A scorecard for dairy inspections is developed. Marker beginning of USDA-regulated farm inspections.

1909 — At W.D. Hoard's prodding, University of Wisconsin is first great university to formally recognize farmers in public tribute. Honored were men who had "stimulated progress, dispelled ignorance and added greatly to the uplift of the farming class."

"There are nothing like facts. *Hoard's Dairyman* is hungry for facts for it realizes that it has a class of readers who live on facts. Opinions are worth nothing unless founded on the conclusion of facts. Sound opinion is the result of a good digestion of sound facts. Alleged facts, or foods, that are made up of moonshine will bring mighty poor digestion."

1910 The 39-year-old company moves into a new building on Milwaukee Avenue in Fort Atkinson, Wis. Still occupies it today. (Shown right)

1912 First discussed loose or pen housing for dairy cattle. Labor-saving advantages and low initial investment stressed through the years. Later accepted nationwide.

"The happiest farmer is he who has time and opportunity to do his work rightly, thoroughly, and most perfectly. There is a keen sense of enjoyment in such work that only the true artist knows."

"If we go among those breeders of dairy cattle who produce the best cows, as told by their after history, we will almost invariably find that they are generous feeders."

1914 The Smith-Lever Act established the Extension Service to bring the fruits of university research directly to the farmer.

1914 W.D. Hoard's portrait hung in famed Saddle and Sirloin Club in recognition of his great contribution to American agriculture.

"The cow has none of the inanimate qualities of the machine. The whole process of elaboration of milk is animal and maternal, not mechanical. The laws of chemistry, physiology, and biology controlled by the animal and her maternal instincts apply, not the laws of mechanics."

1914 First tank trucks used to transport milk from farms to dairy plants.

1915 Began first youth page for boys and girls on dairy farms. Over 75,000 youngsters enrolled in *Hoard's Dairyman* Juniors. Taught fundamentals, encouraged youth involvement. Forerunner of internationally known 4-H and FFA organizations.

"If a dairy farmer is going to get rich in this world, he is going to employ a whole lot of pairs of hands."

"It won't do to handle this magnificent mother (the cow) with a mind that has no study or thought."

1915 National Dairy Council created to educate the public about nutrition.

1915 "Wisconsin's most distinguished citizen" named at World's Fair in San Francisco is W.D. Hoard.

> If the farmers of the United States were as wise on the question of maintaining fertility as they are in the adoption of machinery, what a country this would be.

> When I buy a cow I would rather pay $100 for a cow making 300 pounds of butter than pay $50 for a cow making 250 pounds.

1916 Discovery of vitamins first reported in *Hoard's Dairyman* by Dr. E.V. McCollum.

1916 National Milk Producers Federation formed to represent the dairy farmers in the nation's capitol.

"Naturally, we do not expect everybody to agree with us any more than we expect to agree with everybody, but we hope and expect to cause everybody to think."

"Today, in my own farm work, I feel the necessity of more and more study, for I tell you that it is a more difficult thing to run a dairy farm than it was 40 years ago."

1917 After 22 years of *Hoard's Dairyman* campaigning for tuberculosis eradication, federal government finally launches program. Farmers bootlegged cattle into untested areas. Editors courageously addressed mobs of angry farmers throughout the country.

1917 — *Hoard's Dairyman* was the dominating influence in the founding of the herd test, proved sire, brood cow research program conducted by the U.S. Department of Agriculture.

"A well-equipped dairy farmer must be the best kind of an all-around farmer, for he has to do with all the problems of the soil and crops and especially and specifically with those problems relating to dairy cattle. So, at a glance it can be seen that the true dairy farmer must be the broadest kind of a farmer and the true dairy paper and broadest kind of a farm paper."

1918 November 22 W.D. Hoard passes away.

1918 Arthur J. Glover succeeds deceased founder as editor. Glover pioneered herd production testing, first in U.S., in Illinois in 1902.

"I want to get at the man that produces the milk. I want to see that man's profits enlarged. I want to see his labors lightened. I want to see his intelligence increased. I want to see his family happier and his home more cheerful, and the man and all that belongs to him a better product of this day and civilization. That is what I want to see."

1919 All cattle required to have tuberculin tests prior to interstate shipping.

1922 — A statue of William Dempster Hoard was unveiled on February 3, 1922, at Henry Mall on the University of Wisconsin-Madison campus. The event marked the 50th anniversary of the Wisconsin State Dairymen's Association.

> "If you are determined to let a cow gallop over five acres of land during the summer to get a living, then you need to have very cheap land for her racetrack."

> "The study of the soil is one of the deepest and most important subjects that can claim the attention of mankind, for, without it, the human race could not exist."

1922 Capper-Volstead Act gives legitimacy to agricultural cooperatives by exempting them from antitrust constraints.

1923 — "More milk per acre" articles became school and university texts and exerted profound influence on dairy industry. Pastures were classified as a high-income crop, rather than exercise lots and wasteland.

" We sometimes read about 'heart to heart talks,' but there is no product on earth so intensely 'heartful' as good milk. It is the result of a wonderful union of 'hearts and hands'; the heart of that great mother, the cow, the heart of the owner, the milk, and all the hands that have willingly, thoughtfully, and skillfully ministered to its production. "

1924 *Hoard's Dairyman* campaigned successfully for the elevation of the U.S. Dairy Division to the status of a Bureau in the Department of Agriculture. (Shown right)

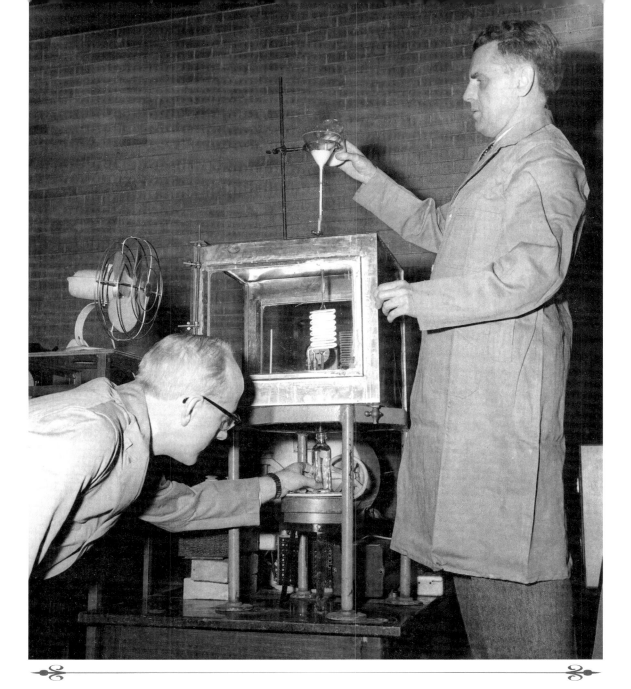

1924 U.S. milk production per cow averaged 4,167 pounds.

"A friend writes us that he believes there is one reason why the advice of *Hoard's Dairyman* is worth what it is. It is because we 'are not so mortal afraid of speaking the truth in a plain way, even if it does shame the devil and lose us a subscriber once in awhile.'"

1925 American Dairy Science Association recommends uniform rules and methods for on-farm milk production testing so records can be compared between states.

1926 Magazine switches from a weekly to a twice-a-month publication. A second color becomes possible with new printing capabilities.

"I know of whole townships of farmers where there has not been $100 spent for dairy knowledge in 10 years, and yet these same farmers wonder that other men make money out of dairy farming."

"The man who extracts out of the business a just degree of profit is the man who applies the largest degree of intelligence and skill."

1928 Circulation had topped 150,000.

1930 — *Hoard's Dairyman* Cow Judging Contest launched. More than 5.5 million entries have been received since it began. Brought art of breeding better dairy cattle into the classrooms of the nation.

"Farmers drift downwards unless they keep their minds actively at work. Very soon they begin to lose financial profit, and they cannot see where the trouble lies. There must be more mind work to save the hand work from actual loss. Men who keep cows must read about cows, think about cows, study about cows, or they will suffer for such lack of reading, thinking, and studying."

1930   A.J. Glover recognized at State Dinner by President Hoover as the man who had made the greatest contribution to American agriculture during the decade.

1933 A.J. Glover elected to serve a four-year term as president of the Holstein-Friesian Association of America now known as Holstein USA.

"You remember the Irishman who said that all whiskey was good but that some whiskey was better than others. It is so with ensilage."

"There is no known combination of forage that can equal alfalfa hay and corn silage. There is a harmony of working between these two feeds that is surprising."

1934 The first refrigerated milk cooler was introduced, enabling large gains in milk keep quality.

1935 There is a nearly record high number of dairy cows on farms... more than 24 million. Average production per cow was 4,184 pounds of milk and 165 pounds of fat. Total national production was slightly over 100 billion pounds.

"It ought to be the object and purpose of an intelligent citizenship to guard well this great dairy industry. Every dweller in the cities should realize that he has a duty to perform, in the defense of the cow and her product, fully as great as is expected of the farmer."

1936 A September 25 editorial on cooperation versus socialism helps form the basis for the magazine's long-standing support of the cooperative movement.

1937 Agricultural Adjustment Act amendment gives Secretary of Agriculture authority to develop milk marketing agreements and create Federal Milk Marketing Order system.

"The work of breeding good animals is a work for a long time, requiring pluck, patience, a high degree of intelligence, and a close practical judgment. Money will buy such animals, but it will not breed them successfully even after it has bought them. That is the work of brains."

1937 The magazine has its first discussion on the advantages of baling hay in the field instead of handling it loose.

1938 First detailed report on artificial insemination in the United States.

"Even the noted Corn Belt of Illinois, Indiana, and southern Michigan, make no such show of corn in advanced and rank growth, as one sees on the farms that are known to be carrying large dairies. The land needs the cow as much as the cow needs the cornstalks."

1938 Readers regularly see articles and advertisements about hybrid seed corn.

1938 First advertisement for a field forage chopper is placed by the Fox River Tractor Co., Appleton, Wis.

"A campaign of education ought to be instituted among the consumers to teach them that milk is the cheapest food at 10 cents a quart that is found on the market and so get them to see that they should be willing to pay the farmer for his extra cost in producing clean, healthy milk."

1938 First article about an elevated milking parlor is in the August 25 issue. The single, four-stalls-in-a-row parlor was on the J. Russell Land Dairy, Grinnell, Iowa.

1939 University of Wisconsin Board of Regents elects A.J. Glover as president, a role he holds until 1943.

"All work requires study, understanding, intelligence. There is no such thing as menial or degrading labor. The work of the farm takes hold on the most profound principles of science. How then can it be ignorant work?"

1939 On November 25, Frank W. Hoard, youngest son of W.D. Hoard, passed away. W.D. Hoard, Jr., grandson of the founder, became president and general manager. (Shown right)

1940 Entire nation declared free from tuberculosis in cattle. Medical leaders hail campaign as "man's greatest victory over tuberculosis." Successful culmination of campaign started by W.D. Hoard 45 years earlier.

" We have never yet seen a woman fall when she took hold of the management of a dairy."

" Happiness doesn't depend on what we have, but it does depend on how we feel towards what we have. We can be happy with little and miserable with much."

1940 There were 4,644,416 farms with milk cows.

1940 First advertisement for a mechanical barn cleaner appears in the February 10 issue.

"A cow of dairy blood and performance eats largely in order that she may turn out large quantities of butter or cheese. With a thoroughly good cow to manufacture it, we can always afford to put in feed and take out butter. That is what a good dairyman is here for."

1943 The first uniform scorecard for type is approved by the Purebred Dairy Cattle Association and the American Dairy Science Association.

1944 There are 25,661,000 dairy cows on U.S. farms, the greatest number in history.

"When dairymen get right down to bottom facts they will find a silo, with a supply of well-cured silage in it for cow feed in the hot, dry weather of summer, a paying investment. I had 30 tons of silage left last spring, which I fed this summer and now have no doubt about its value as a summer feed."

1946 Experimental silo unloader is being tested on several Midwest dairy farms.

1946 *Hoard's Dairyman* supports renewed campaign against brucellosis, most serious livestock menace. Livestock industry confused, torn by dissension. Incidence doubled since 1941. Disease causes undulant fever in humans.

" There is much more to this matter of breeding than the mere mating of animals. It has to do with the deepest laws of animal physiology and every breeder should endeavor to be guided by the facts as they have been worked out by eminent students. But even then the real successful breeder is a man with a born love and adaptation to the work."

1948 First national magazine column on artificial breeding of dairy cattle. Greatest single development in dairying since origin of herd test in 1902.

1948 J.S. Baird (1948), E.C. Meyer (1948) and W.D. Knox (1941) begin careers with *Hoard's Dairyman* during the decade. Together they gave 183 years of service to the dairy industry.

"We don't need any more lawyers or any more bankers, but God knows we do want more intelligent farmers, and we never can get them until we commence to train the young mind as they do in Europe."

1949 A.J. Glover, the second editor of the magazine, dies on May 8 at age 76. He was succeeded by W.D. Knox, a Michigan native, who had been a member of the editorial staff since 1941. (Shown right)

1949 Uniform national program approved for brucellosis control. *Hoard's Dairyman* editor, W.D. Knox, elected to lead medical, farm and dairy industry forces to eradication goal through National Brucellosis Committee.

"Prevention of udder troubles is much better than hunting for medicine to cure them. No matter what kind of stall you use, see to it rigorously that the cow has a warm bed to lie on, and you will about entirely escape all udder difficulties."

"We are coming rapidly to understand that we have got to furnish pure air to our cows."

1949 Pipeline milking is tried at Excelsior Dairy at Santa Ana, Calif.

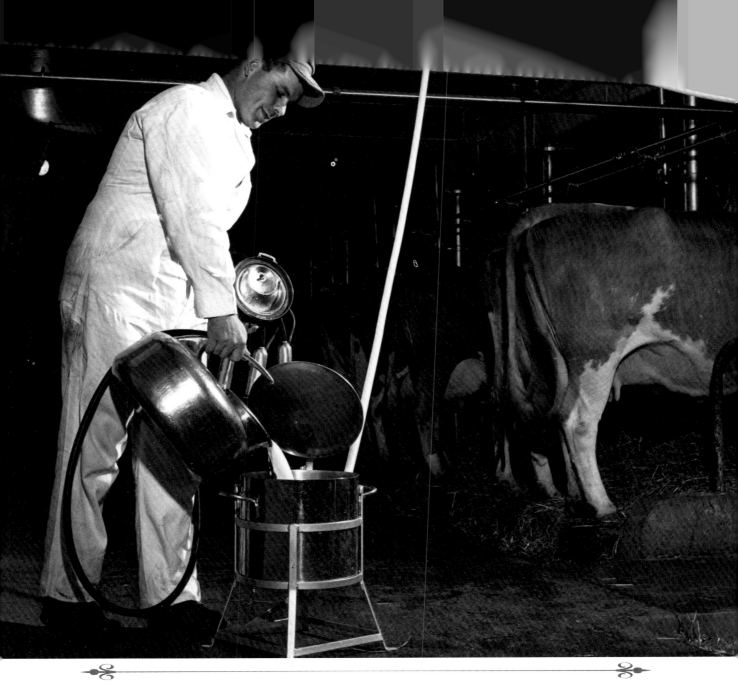

1951 *Hoard's Dairyman* Plan Service established with introduction of a 25-cow, two-story stall barn.

" Remember that a cow is a mother, and her calf is a baby."

" The secret of success in handling cows is an understanding of the laws and functions of motherhood."

" I tell you this, as true as there is a God in heaven, a heifer is made or unmade the first year of her life."

1951 Tense and divided Wisconsin livestock industry asks *Hoard's Dairyman* editor, W.D. Knox, to lead effort to develop effective disease control program. Five years later, state is free of brucellosis and Wisconsin is accepted model for all states.

1951 First calf is born as a result of embryo transfer . . . termed "incubator" calf.

> "If we were to name any one thing more necessary than all the rest to keep calves healthy and free from scours it would be this: see to it that the calf is kept constantly dry and clean."

> "The lack of profit this very year on thousands of dairy farms, is not so much the high price of feed as it is stupidity of keeping a lot of cows that are utterly unfit for the business."

1951 The first commercial milk replacers for dairy calves were introduced to the marketplace.

1953 Famed Ed and Emma cartoon series debuts in *Hoard's Dairyman*. Drawn by Chuck Stiles, Ed and Emma is the longest continuous cartoon series drawn by one person.

"I never look at a cow but that I think of her with humility and a feeling of awe and inspiration."

"The only way to secure a tolerable degree of certainty in dairy qualities is to breed from thoroughbred dairy bulls."

1953 Postwar milk production begins to outstrip demand and dairy prices falter. *Hoard's Dairyman* editorial, "Hour of decision—sell or suffer," starts nationwide voluntary contributions for advertising dairy products, research and market development.

1953 Five daughters of Play Haven's Y. Plymouth of the Hoard's Dairyman Farm classify Excellent.

" The formation of the udder is indicative of the talent of the cow, indicative of her purpose, indicative of her ability. It should rise high in the rear and extend well forward upon the abdomen."

" If you are keeping 20 cows, and you have 7 or 8 or 10 that are poor ones, get rid of them and keep 10 that will make a profit."

1953 Frosty, the first calf from frozen semen in the U.S., born on May 29.

1954 Postwar buildup of farm surpluses reaches critical stage. *Hoard's Dairyman* advances self-help dairy program designed to rid government of heavy costs and surpluses and place program in hands of farmers for self-financed market stabilization.

"I believe that one of the most important things for the farmer is to cultivate a manly pride in his business, pride in his progress, pride in his boys, pride in his girls, pride in his cattle, pride in his name, and pride in his honor as a farmer."

1956 — *Hoard's Dairyman* Continuing Market Study is launched. Annual survey of 3,000 readers gives most in-depth look into the dairy industry.

1956 The red title block that has become the magazine's trademark is added.

"The cow is the 'Foster Mother of the Human Race' - from the day of the ancient hindoo to this time have the thoughts of men turned to this kindly and beneficent creature as one of the chief sustaining forces of human life."

1957 New yardstick of dairy efficiency advanced by *Hoard's Dairyman* to guide nation's dairy farmers. More-milk-per-man concept is first benchmark to adequately relate labor efficiency to dairy profit.

1957 Jim Baird paints the first "Foster Mothers of the Human Race" painting based on cows photographed at National Dairy Cattle Congress in Waterloo, Iowa.